Counting It Up

by Michele Koomen

Consultant:
Deborah S. Ermoian
Mathematics Faculty
Phoenix College
Phoenix, Arizona

Bridgestone Books
an imprint of Capstone Press
Mankato, Minnesota

Bridgestone Books are published by Capstone Press
151 Good Counsel Drive, P.O. Box 669, Mankato, Minnesota 56002
http://www.capstone-press.com

Library of Congress Cataloging-in-Publication Data
Koomen, Michele.
    Numbers: counting it up/by Michele Koomen.
    p. cm.—(Exploring math)
    Includes bibliographical references and index.
    ISBN 0-7368-0818-3
    1. Counting—Juvenile literature. [1. Counting.] I. Title. II. Series.
QA113 .K66 2001
513.2'11—dc21   00-010565

Summary: Simple text, photographs, and illustrations introduce numbers and counting,
including place value, comparing numbers, and examples of how people use numbers.

Editorial Credits
Tom Adamson, editor; Lois Wallentine, product planning editor; Linda Clavel, designer;
    Katy Kudela, photo researcher

Photo Credits
Artville/Jeff Burke and Lorraine Triolo, 14 (all), 15 (all)
Capstone Press/CG Book Printers, cover, 12
Image Farm Inc., 16 (upper right), 16 (lower right)
Kimberly Danger, 18
Photri-Microstock, 16 (upper left)
Shaffer Photography/James L. Shaffer, 20
Visuals Unlimited/Tom Edwards, 16 (lower left)

1 2 3 4 5 6 06 05 04 03 02 01

Table of Contents

Numbers and Counting

Numbers tell how many objects are in a group. Counting is giving a number to each object to find the total. Count these flowers.

1

2

3

4

5

7

6

8

The first flower is "1." The second flower is "2," and so on. The last number will tell you how many flowers are in the group. There are 8 flowers.

6

Tens and Ones

This group has many rocks. You could count them one by one. But there is an easier way to count this many objects. You can group the rocks into tens and ones.

10

10

3

10 + 10 + 3 = 23

Counting by Tens and Ones

Count 10 rocks. Put them in a pile. Count 10 more rocks. Put them in another pile. There are 3 rocks left over. You have 2 groups of ten and 3 ones. Together, they make 23.

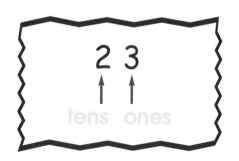

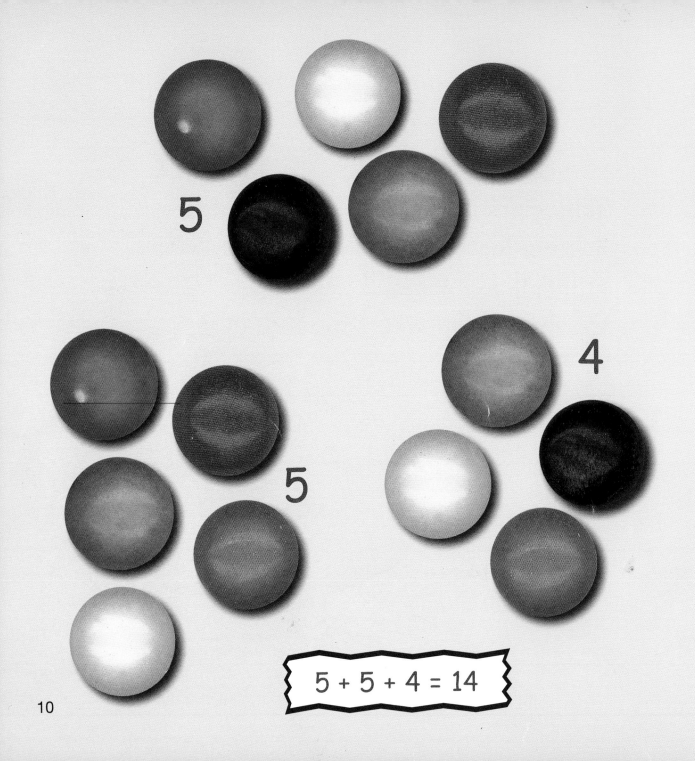

5

4

5

5 + 5 + 4 = 14

Tally Marks

Tally marks help you keep your place in counting. Make a mark for each gumball you count.

Make a diagonal mark when you get to 5. Grouping the gumballs by 5s will make it easier to count the total.

There are 14 gumballs in the group.

Comparing Numbers

Numbers help us to compare how many objects are in different groups. Compare each group of toys in the picture. One group has more toys. One group has fewer toys. The number of toy cars is greater than the number of yo-yos.

This group has more cherries than cucumbers. There are 7 cherries and 4 cucumbers. The number 7 is greater than the number 4.

7 > 4

↑

This sign means "greater than."

This group has fewer pineapples than strawberries. There are 2 pineapples and 4 strawberries. The number 2 is less than the number 4.

2 < 4

↑

This sign means "less than."

Stovepipe Wells	22 mi	35 km
Jct (136)	83 mi	134 km
Jct 395	97 mi	156 km

SPEED LIMIT 30

TO

INTERSTATE
GEORGIA
185

→

16

MAXIMUM
40
km/h

Numbers on a Trip

Numbers give us information when we go on a trip. Numbers tell us which road to take. Numbers tell us how fast we can drive a car on the road. Numbers also tell us how many miles or kilometers we need to travel to get to a place.

Numbers and Age

Every year you celebrate your birthday. You use a number to tell other people your age. Your age tells the number of years it has been since you were born. Count the candles on this birthday cake to see how old this boy is.

HOME **15** 7:02 TIME *Naden* VISITOR **6**

2 QTR **4** DOWN **01** TO-GO

20

Numbers and Sports

Numbers tell us information when we play sports. A scorekeeper counts all the points scored in a game. The score is the number of points that each team earns. The team that earns the highest number of points wins. In what other ways do people use numbers?

Hands On: Number Hunt

This game will help you practice counting. You can play either in teams or by yourself.

What You Need

Paper
Pencil or pen
Several players

○	1
	2 doors
	3
	4 pieces of chalk
○	5
	6
	7
	8
	9
○	10

What You Do

1. Number your paper from 1 to 10.
2. In your classroom or in your house, find a group of objects for each number. For example, find 4 pieces of chalk, or find 2 doors.
3. Do not count the same objects for more than one number.
4. The first player to fill all 10 lines wins.

Make up different rules for this game. Who can find the largest group of objects? Who can find the most colorful objects? See if you can find the groups of objects in order.

Words to Know

celebrate (SEL-uh-brate)—to do something fun on a special occasion

compare (kuhm-PAIR)—to point out how things are alike or different

diagonal (dye-AG-uh-nuhl)—a line that is at an angle

information (in-fur-MAY-shuhn)—facts and knowledge

Read More

Cato, Sheila. *Counting and Numbers.* A Question of Math Book. Minneapolis: Carolrhoda Books, 1999.

Demi. *One Grain of Rice.* New York: Scholastic, 1997.

Patilla, Peter. *Numbers.* Math Links. Des Plaines, Ill.: Heinemann Library, 2000.

Schmandt-Besserat, Denise. *The History of Counting.* New York: Morrow Junior Books, 1999.

Internet Sites

Ask Dr. Math
http://mathforum.com/dr.math
Aunty Math
http://www.dcmrats.org/auntymath.html
Plane Math Activities
http://www.planemath.com/activities/pmactivities4.html

Index